EXPLORING THE SKY

THE EQUINOX GUIDE TO WEATHER AND THE ATMOSPHERE

TERENCE DICKINSON

FIREFLY BOOKS

A Firefly Book

Published by Firefly Books Ltd.

Copyright © 1988, 1995 Terence Dickinson

First published in 1988 by Camden House Publishing
(a division of Telemedia Communications Inc.)

Eighth printing 2009

Publisher Cataloging-in-Publication Data (U.S.)
(Library of Congress Standards)

Dickinson, Terence.
  Exploring the sky by day : the Equinox guide to weather and the atmosphere / Terence Dickinson; principal illustrations by John Bianchi.

[72] p. : col. ill., photos. ;   cm.

Includes index.

Summary: A guide to weather and atmospheric phenomena from cloud formations and sundogs to lightning and tornadoes.

ISBN-13: 978-0-92065-671-6 (pbk.)
ISBN-10: 0-92065-671-4 (pbk.)
ISBN-13: 978-0-92065-673-0 (bound)
ISBN-10: 0-92065-673-0 (bound)

1. Weather. 2. Climatology. I. Bianchi, John. II. Title.
551.6   21   QC863.4.D53   1988

National Library of Canada Cataloguing in Publication Data

Dickinson, Terence
  Exploring the sky by day

Includes index.
ISBN-13: 978-0-92065-671-6 (pbk.)
ISBN-10: 0-92065-671-4 (pbk.)
ISBN-13: 978-0-92065-673-0 (bound)
ISBN-10: 0-92065-673-0 (bound)

1. Meteorology – Juvenile literature. I. Title.
QC863.5.D52   1988   j551.5   C88-094431-5

Published in the United States by
Firefly Books (U.S.) Inc.
P.O. Box 1338, Ellicott Station
Buffalo, New York 14205

Published in Canada by
Firefly Books Ltd.
66 Leek Crescent
Richmond Hill, Ontario L4B 1H1

Design by
Ulrike Bender

Printed and bound in Canada by
Friesens
Altona, Manitoba

Printed on acid-free paper

Front Cover: Solar halo and sundogs. Illustration by John Bianchi (details, page 46).

Back Cover: Waterspout over Lake Winnipeg. Photograph by John Scott Norquay (details, page 64).

OTHER BOOKS BY TERENCE DICKINSON
NightWatch (3rd Edition)
The Universe and Beyond
Exploring the Sky by Day
The Backyard Astronomer's Guide, 2nd Edition
  (with Alan Dyer)
From the Big Bang to Planet X
Extraterrestrials
Other Worlds
Summer Stargazing
Splendors of the Universe (with Jack Newton)

CONTENTS

INTRODUCTION

The sky is a constant reminder of both the power and the beauty of nature. The atmosphere that surrounds Earth is a complex weather machine—fascinating to watch and, occasionally, a brutal experience to suffer.

While we have the technology to observe storms and to predict the weather, we cannot control it. Even in our advanced scientific age, tornadoes still kill more than 100 people a year in North America, and when a hurricane blows in off the Atlantic Ocean, we are powerless to prevent it from destroying homes and flooding towns. A blizzard can still shut down a city for days at a time, just as it did 50 years ago. And long droughts can destroy an entire nation's economy, leading to starvation for tens of thousands of people who can only wait helplessly for rain.

But the sky is more than just a place where disaster brews. It is an important part of our environment that can be studied and enjoyed as much as forests and fields and oceans. The day sky is our window on the weather and even on space, where the sun is millions of times more distant than the outer edge of the Earth's atmosphere. Our atmosphere shields us from the sun's harmful rays, acts as a buffer from the lifeless vacuum of space and, at the same time, gives us the colours of the sky.

Weather of all kinds can be observed by everyone without any special equipment. We can watch storms approaching and appreciate why changes in weather occur. We can classify clouds and interpret their meaning. We can understand rainbows and measure the distance to lightning bolts.

The sky is alive, changing constantly, often before our eyes. All we need do is look up to enjoy the glory of one of nature's most fascinating and accessible phenomena.

5

1 THE ATMOSPHERE

Think of the atmosphere as an ocean of air. Just as fish swim in oceans of water, birds travel in an "ocean" of air. We live at the bottom of that same air ocean. But while the ocean's water is a liquid, the atmosphere's air is a gas. Actually, it is a mixture of gases—a mixture that is pretty much the same everywhere on Earth.

Suppose we could separate the air in an average living room into layers of each different gas. Just over three-quarters of the room (78 percent, to be exact), from the floor to the top of an adult man's head, would be filled with nitrogen. Humans do not need nitrogen, but it mixes with soil and provides energy for plants.

Most of the rest of the living room would be oxygen (21 percent). Oxygen is an essential fuel for animal life. As we breathe, oxygen from the air is absorbed into the blood through our lungs.

Oxygen and nitrogen together make up 99 percent of the atmosphere. Only the top two to three centimetres (about one inch) of our room is left for other gases. These are called trace gases. They include methane, neon, helium, hydrogen, ozone, argon and, most important, carbon dioxide, which plants absorb and use for growth.

Water vapour is a gas in the atmosphere, too, but it changes so rapidly between gas, liquid and solid states that it is not considered a permanent atmospheric gas.

As shown in photographs like this, taken from a great distance, the atmosphere hugs Earth like an invisible skin. Only the clouds are visible. The air itself is transparent.

The atmosphere protects Earth. Although thin and transparent, the air blanket shields us from cosmic rays and other dangerous radiation from the sun and deep space. It is an insulator, too, providing protection from the extreme temperatures of space. Without it, almost every place on Earth would be either blisteringly hot or frigidly cold, as is the case on the moon.

This photograph of Earth was taken on December 7, 1972, by the Apollo 17 astronauts on their way to the moon. At the top of the globe are the Sahara and Arabian deserts, seen as mostly brown. The Indian Ocean is at the right, the Atlantic at the left, and Africa is between them. Ice-covered Antarctica is at the bottom, surrounded by clouds.

The splendour of the sky is there for all to see on hundreds of days each year. Anyone can become a knowledgeable skywatcher.

The atmosphere from top to bottom is several hundred kilometres deep. Like everything else on or near the Earth's surface, the atoms and molecules in the air are prevented by gravity from hurtling into space. (Some hydrogen, the lightest gas, does get away, but all of the other gases are trapped.)

The Earth's gravity pulls the atmosphere's atoms and molecules so effectively that most of them stay within a few kilometres of the ground. That means the air is thickest near the Earth's surface and gets thinner with increasing altitude. At the tops of the Earth's tallest mountains, the air is too thin for us to breathe.

The lowest, densest zone of the atmosphere is the troposphere. It could also be called the weather zone, since almost all clouds, rain and snow occur here.

Above the troposphere is the stratosphere, where the air is so thin that only specially designed airplanes can fly in it. Beyond the stratosphere, it is almost like outer space, but there are still enough air particles to protect us from meteors, cosmic rays and other uninvited guests from deep space. Eventually, though, the air particles thin out and the true vacuum of space begins.

Every day, Earth runs into millions of bits of material (rocks and icy chunks) left over from the formation of the planets. When this debris strikes air particles at speeds of up to 240,000 kilometres per hour (150,000 mph), it burns up from friction. The incineration is seen as a bright flash— a meteor. Meteors are sometimes called falling stars or shooting stars, but they have nothing to do with stars at all.

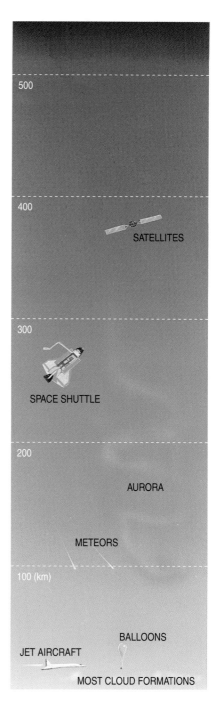

500

400

SATELLITES

300

SPACE SHUTTLE

200

AURORA

METEORS

100 (km)

JET AIRCRAFT

BALLOONS

MOST CLOUD FORMATIONS

2 FORMATION OF CLOUDS

- What are clouds made of?
- How are clouds born?
- How much does a cloud weigh?
- Can clouds form at ground level?

△ You can't touch a cloud, and if you see one up close, it is merely a fog— countless tiny droplets of water. In fact, a fog is a cloud that forms at ground level. On a cold day, a person's breath makes a miniature cloud exactly like the ones in the sky. The moisture (water vapour) in the warm air from the lungs comes into contact with the cold outside air and condenses into water droplets. The same thing happens to water vapour escaping from a pot of boiling water: the cooler room air causes condensation, and a tiny cloud develops.

All clouds are water in one form or another, and it does not take much water to make a cloud. A small cloud the size of a school playground contains less water than a full bathtub and would weigh about as much as an adult man. The water that makes up the big, billowing cloud pictured here would probably fill several thousand Olympic-sized swimming pools.

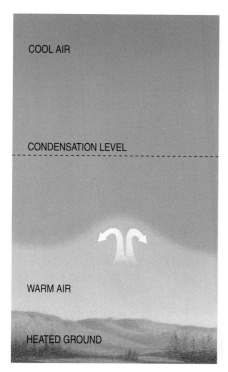

COOL AIR

CONDENSATION LEVEL

WARM AIR

HEATED GROUND

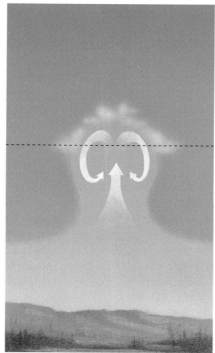

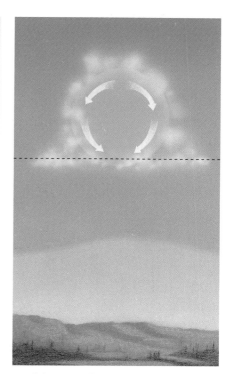

△ The story of cloud formation begins at ground level. Sunlight heats the ground, and the warm ground heats the air next to it. The hot air rises. A parcel of air does not have to be much hotter than the air around it—just a degree or two—to start slowly rising.

Since some sections of the ground are grass and some are trees or houses or ponds, the air over them heats unevenly. Certain pockets of air rise faster than others. How a pocket of rising air forms a cloud is shown above.

Air at ground level is denser than air higher up. So as an air parcel rises, it expands and becomes less dense. But expansion cools the air. As it cools, some of the water that is present in the air as vapour (gas) condenses into minuscule droplets that are far too small to be seen with the eye alone.

Initially, the droplets form around tiny particles in the air (usually dust or pollen) called condensation nuclei. The droplets gather in incalculable numbers to form clouds. A cloud is therefore a big blob of extremely tiny droplets of water, or it can be frozen water in the form of small ice crystals or snowflakes.

But why do clouds form on some days and not on others? Cloud formation depends on many factors, but the most important is the amount of water in the atmosphere. If there is not enough water, there will be no condensation into droplets and no clouds.

Clouds of major weather systems, which are overhead for several days in a row, take much longer to form than the type shown above. As a large mass of air moves, forces in the atmosphere can cause it to rise and cool. This can result in the formation of a blanket of cloud that can be hundreds of kilometres across.

The atmosphere gets its water in the form of vapour from oceans, lakes and plants. Just as water in a bowl will disappear in a day or two by evaporating, water in lakes and oceans evaporates into the air. Plant roots draw water from the ground, and most of it evaporates through the plant's leaves.

3 TYPES OF CLOUDS

- What are the basic kinds of clouds?
- What do the different types look like?
- Why are there different clouds?
- When were clouds named?
- How high in the sky are the clouds?

▷ Two hundred years ago, the clouds had no names. Big puffy clouds, thin streaky clouds and dull rain clouds were all just clouds. Nobody knew how they formed or why different types of clouds occurred.

Luke Howard, an English pharmacist who spent his spare time studying nature, decided in 1803 that he would name the various cloud types so that people could properly describe and study them. The system he developed is still being used.

Howard had noticed that all clouds belong to one of three basic groups. The puffy clouds often seen on a summer afternoon he called cumulus, which means "heap" or "pile" in Latin. The second distinct family is made up of wispy high-level clouds. Howard named them cirrus, the Latin word for "curl." The third group in his classification, the clouds that lie in great horizontal blankets, he called stratus, which means "stretched out."

Howard realized that the three categories covered only the primary cloud types, so he added some names that could be used in combination with the others to describe the clouds more accurately. Today, 10 different cloud classifications based on Howard's system are recognized by skywatchers around the world.

At the highest levels, there are cirrus, cirrostratus and cirrocumulus clouds. Midlevel clouds are altostratus and altocumulus. Low-level clouds are stratus, nimbostratus and stratocumulus. The two remaining types are sometimes so tall that they can be low-, medium- and high-level clouds all at the same time. These are cumulus and cumulonimbus. Here is a brief description of each of the 10 types:

Cirrus are beautiful, feathery clouds, sometimes called mare's tails because they resemble a horse's tail blowing in the breeze.

Cirrostratus clouds are white or pale grey veils that can extend over part or all of the sky. The sun is easily visible through these thin clouds.

Cirrocumulus clouds are white or pale grey with a definite structure—usually ripples or ribs. The sun is visible through them.

Altostratus are formless, greyish clouds that can cover part of the sky or extend from horizon to horizon in a great sheet. The sun is visible through altostratus.

Altocumulus can cover all or just part of the sky. They are patterned clouds, usually rippled. Sunlight penetrates altocumulus fairly well.

Stratus are dull grey clouds that cover the whole sky and often completely block direct views of the sun. These clouds can begin as a fog bank that rises.

Nimbostratus are dark grey clouds that usually mean a dull day and possibly rain. They completely blanket the sky and have little visible structure. Rain from nimbostratus clouds can last for a day or more.

Stratocumulus clouds cover the sky, but unlike the formless altostratus, they have definite bumps, humps and rolls. These thick clouds rarely let the sun peek through.

Cumulus clouds are white, puffy clouds shaped like giant cauliflowers. They are most often seen on summer afternoons.

Cumulonimbus are storm clouds that can produce sudden, violent weather.

Clouds change from one type to another when moisture or temperature variations occur. So when you see a change in cloud type, a change in the weather is under way.

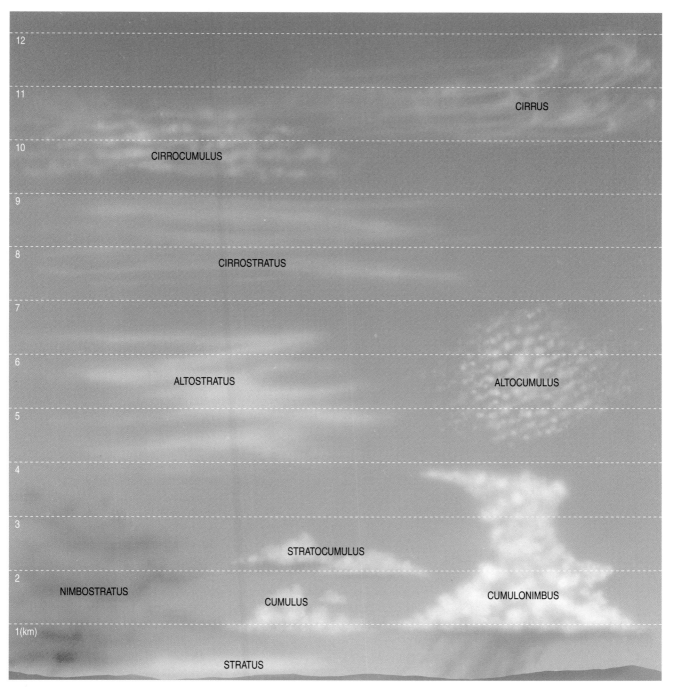

12

11 · CIRRUS

10 · CIRROCUMULUS

9

8 · CIRROSTRATUS

7

6 · ALTOSTRATUS · ALTOCUMULUS

5

4

3

STRATOCUMULUS

2

NIMBOSTRATUS · CUMULUS · CUMULONIMBUS

1(km)

STRATUS

4 CUMULUS CLOUDS

- Why are cumulus clouds so puffy?
- Why are some cumulus clouds pure white, while others are greyish?
- Why are there different types of cumulus clouds?
- What happens inside a cumulus cloud?

△ You can watch puffy white cumulus clouds, like the ones in these photographs, form on a sunny summer afternoon when the sun has warmed the ground and huge bubbles of heated air rise into the sky. The rising air soon cools, and some of its water vapour condenses into a cumulus cloud. The cloud is made up of tiny water droplets which reflect sunlight so well—almost as well as snow does—that the cloud appears white. Thick cumulus clouds appear darker on the bottom because sunlight is scattered so much it is partially blocked.

Cumulus clouds are rarely seen in winter because it is too cold for the cloud droplets to remain liquid. Also, snow-covered ground does not radiate enough heat to get much warm air rising. Cumulus clouds are, therefore, warm-weather clouds.

Observing cumulus clouds is the most enjoyable cloud-watching activity. They are always changing in size and shape. Some emerge from nowhere into the blue sky. Others shrink and disappear, especially later in the day when the sun is low in the sky and the ground begins to cool.

◁ Rising air currents sometimes create huge cumulus clouds that look like gigantic blossoms in the sky. Watch the tops and upper edges of these clouds. They constantly change, growing here, shrinking there. Yet no matter how huge a cumulus cloud becomes, it always has a flat bottom. The bottom marks the level above the ground where the air is cool enough for water vapour to condense into cloud droplets.

Inside a cumulus cloud are complex updrafts and downdrafts that carry some parts of the cloud higher while pulling other parts lower. In a big, growing cumulus, there are more updrafts, and hence the cloud builds higher and higher.

Pilots try to avoid flying through cumulus clouds because the unpredictable drafts can push a plane up and down roughly. Airplane crews insist that passengers wear their seat belts when these clouds are around. But from the ground, cumulus clouds are among nature's most beautiful creations.

▷ The mighty cumulonimbus cloud is the biggest, most impressive cloud of all. Such clouds usually form in late afternoon on a hot day when upward currents of warm air rise to their greatest heights. The currents produce the giant tower of cloud that characterizes the great cumulonimbus.

Foul weather brews inside a cumulonimbus, however. Thunder, lightning, heavy downpours of rain—even hail—emerge from the giant storm factories. During the life cycle of a thunderstorm (one or two hours), a single huge cumulonimbus cloud releases the same amount of energy as does the explosion of an atomic bomb.

5 STRATUS CLOUDS

- What are stratus clouds?
- How do stratus clouds form?
- What are they like inside?
- Why do they bring dull weather, sometimes for several days in a row?

△ Stratus clouds always blanket the sky. Sometimes, the sun is faintly visible through them; at other times, stratus clouds totally block the sun, making the day dull and grey.

Pictured above are the prettiest stratus clouds, stratocumulus. Like cumulus, they have a bit of fluffiness but are basically flat sheets with a few openings for blue sky to peek through.

Stratus clouds occur when a large mass of air slowly rises and moisture gradually condenses as the air cools. Stratus clouds may form when cool air moves into an area near ground level and pushes a warmer air mass upward.

Altostratus clouds, not shown on these pages, are seldom noticed because they are just a greyish veil over the sky, partly hiding the sun but never causing foul weather. Altostratus and cirrostratus are mostly composed of tiny water droplets and, depending on the cloud's height and the surrounding air temperature, some ice crystals. These two types of stratus clouds often mark the outskirts of a large region of rain or snow, as illustrated on the facing page.

 Typical stratus clouds, such as those shown here, are generally seen on dull, overcast days. They release rain and snow showers or just give cloudy weather for hours or days at a stretch.

 Nimbostratus clouds are rain and snow clouds. They are low, thick, dark grey clouds in a monotonous cloak overhead and are made up of droplets, raindrops, snowflakes and ice crystals. Long, heavy rains and snowstorms come from nimbostratus clouds because they usually extend over a large area. It just keeps raining until the clouds move on or dump their load of moisture.

6 CIRRUS CLOUDS

- What are cirrus clouds?
- What are cirrus clouds made of?
- What are mare's tails?
- What is a mackerel sky?

Cirrocumulus clouds are the third in the family of three types of cirrus clouds. They are definitely rippled compared with the feathery appearance of cirrus and the featureless veil of cirrostratus. Cirrocumulus clouds sometimes form a pattern called mackerel sky, because the ripples can look like the scales of a fish.

Cirrus clouds are high-altitude clouds composed of multitudes of minute ice crystals. They are always more than five kilometres (3 miles) above Earth, where the temperature is well below freezing even in summer. Cirrus clouds are often the first type of cloud to appear in a clear blue sky.

True cirrus clouds are pure white and wispy, as in the photograph on the facing page. Because they often look like feathers or the swishing tails of horses on the run, cirrus clouds are also called mare's tails. Cirrus clouds have this appearance because after they form, the tiny ice crystals that make up the clouds are carried by winds as they slowly fall. Changing wind currents at different altitudes tend to twist and spread the cloud crystals into wonderful feathery streamers.

Cirrostratus clouds, not shown on these two pages, are pale milky clouds with no structure. They are like a fine veil over the sky. Made entirely of ice crystals, cirrostratus can cause some interesting "special effects" in the sky, as described on page 46. A cirrostratus cloud veil following cirrus clouds is a reliable sign that the weather is changing for the worse. Stratus clouds are probably on the way.

7 UNUSUAL CLOUDS

- What causes cloud formations that are different from the 10 main classifications?

- Where in Canada and the United States can unusual clouds be seen?

Getting to know the 10 main cloud types is the first step in becoming an accomplished skywatcher. However, in mountainous regions such as the Rockies in North America and the Alps in Europe, there is a common type of cloud, called a lenticular cloud, not seen in relatively flat countryside.

Lenticular clouds form only over mountain peaks. Sometimes, they look like a stack of giant dinner plates or a formation of huge flying saucers. They can also look like a huge lid or a tall hat hovering above the mountains. The cause is quite simple. Winds carrying air over a mountain rise up one side, cooling on the way. Moisture in the air condenses to form a cloud. When the air moves down the other side of the mountain, it warms up. The droplets in the cloud then turn back to water vapour because warm air can hold more vapour than cool air.

Since winds are constantly blowing over the mountain, the cloud is continually renewed. It therefore appears to remain in position, since it marks the spot where air is cooled into condensation. Lenticular clouds are not always seen over mountains. The air flow must be smooth, rather than turbulent, to allow the beautiful layer-cake structure to develop.

△ Twenty to thirty times each winter in southern Alberta and in parts of Montana and Colorado, a warm wind arrives that can raise the temperature by as much as 25 Celsius degrees (45 Fahrenheit degrees) in one hour. The wind, a chinook, produces the special lenticular cloud seen in this photograph.

Chinooks originate over the Pacific Ocean waters off the coast of British Columbia and Washington State. Air over the Pacific is much warmer than air over Alberta, Montana and Colorado in winter. When that warm Pacific air rises over the mountains, it creates clouds in the same way lenticular clouds are formed. Then the air moves down the other side of the mountain, heating up again as it descends, and pushes a few hundred kilometres over the prairies as a chinook. The rush of warm air can melt snow in just a few hours and turn winter into a temporary spring.

Strong chinook winds can be unwelcome too. During the 1988 Winter Olympics in Calgary, the ski jumps had to be closed for several days. The strong winds also blew dust and gravel onto the bobsled run and other ice areas.

19

TYPES OF PRECIPITATION

- Why does the water in the clouds start to fall?
- Why does rain freeze sometimes?
- What is the difference between snow, sleet, hail and freezing rain?

When we listen to the weather reports on the radio or watch the weather forecaster on television, the main thing we want to know is, Will it rain or snow today? Precipitation is important to everybody. People need to know what to wear, what the driving conditions will be, whether to take an umbrella, and so on.

The atmosphere is the great recycler, carrying water from one place to another and then dumping it as precipitation. It is the key link in the chain of events that allows living things to exist almost everywhere on the planet.

Even when there are no clouds, there is water in the atmosphere. It is an invisible gas in the air around us. But the gas can change to liquid and fall to Earth as raindrops. Or it can be solid, as snow and ice. All of these forms of water falling from the sky are precipitation.

The first step in the precipitation cycle is the formation of the clouds themselves. The newly formed clouds contain water droplets that are, on average, too small to be seen without a microscope. Even though the temperature of the air in a cloud is generally well below freezing, the droplets are so small that they usually do not freeze right away. In this condition, they are said to be supercooled.

Sooner or later, though, one does freeze to become a tiny ice crystal. Immediately, the neighbouring droplets are attracted to the crystal and merge with it. When a crystal grows larger and heavier, the force of gravity pulls it downward. As it falls, it encounters more droplets that add to the crystal's size. A snowflake is born.

Depending on the temperature above the ground and close to the ground, the precipitation can be snow, freezing rain, sleet (ice pellets), rain or hail, as shown in the diagram.

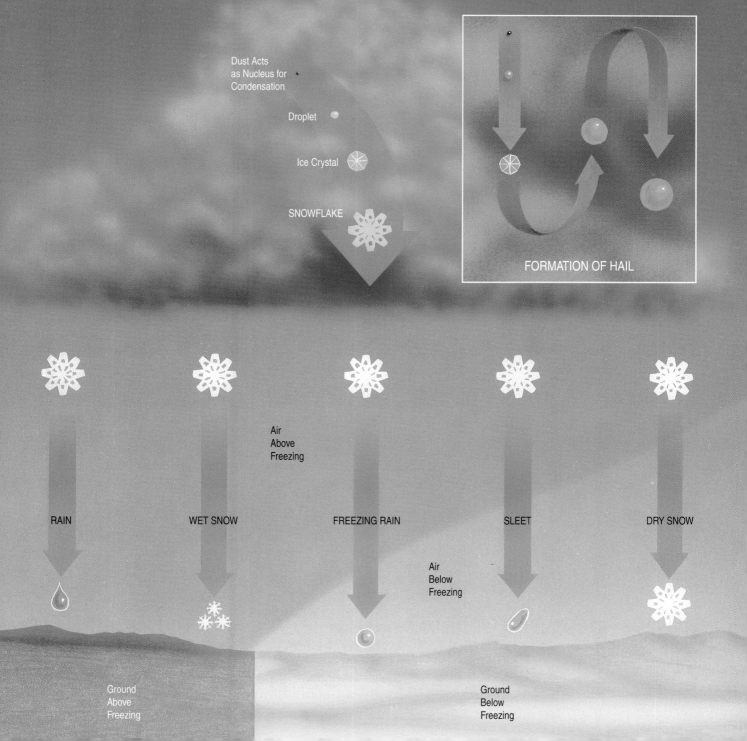

Dust Acts
as Nucleus for
Condensation

Droplet

Ice Crystal

SNOWFLAKE

FORMATION OF HAIL

Air
Above
Freezing

RAIN

WET SNOW

FREEZING RAIN

SLEET

DRY SNOW

Air
Below
Freezing

Ground
Above
Freezing

Ground
Below
Freezing

9 RAIN FORMATION

- How much rain does a cloud hold?

- How big are raindrops before they fall from a cloud?

- Does rain fall from the middle of a cloud or the bottom edge?

- Why are some rainstorms just showers, while others are downpours?

- How fast do raindrops travel?

 A cumulonimbus rain cloud like the one in the photograph above seems to be a giant sponge in the sky—a great floating reservoir of water that, when full, somehow squeezes out its moisture, dumping it on the landscape. But it doesn't work quite that way. As dark as a rain cloud looks, it is hardly "full" of water. Although this cloud could drop enough rain to fill 10,000 swimming pools before it disperses, that is not much water compared with the cloud's size, which is equal to a small mountain.

The tiny droplets that form on dust particles to create a cloud are actually as far apart from one another, in proportion to their size, as baseball players are when they stand at their positions in the outfield. Yet there are so many billions of them that they look like a solid mass.

A droplet grows as more moisture from the air condenses on its surface. When it becomes almost as big as the period at the end of this sentence—a size we would call mist—it begins a slow descent from the cloud interior toward the bottom of the cloud. Still growing, the mist droplet eventually becomes slightly larger than a period and begins falling faster. If there is no wind and little turbulence within the cloud, the droplet will remain this size and fall to the ground as light rain or drizzle. Because of cold temperatures at the altitude of clouds, droplets often freeze and then melt on their way down.

 Small raindrops fall from stratus and nimbostratus clouds like those pictured on page 15. Larger raindrops form in towering cumulonimbus clouds like those seen here. Updrafts and currents in the huge clouds carry drops up into colder air that turns them into snowflakes. This photograph clearly shows the turbulence caused by strong currents within a cumulonimbus cloud. In 1959, a pilot had to bail out of a disabled aircraft into a cumulonimbus storm cloud. His parachute was blown up, down and around inside the cloud for almost an hour before he finally descended to the ground.

Snowflakes inside a cumulonimbus cloud bump together and stick in clumps. When downward currents in the cloud eventually carry the flakes to warmer levels, they melt into fat drops that plunge to Earth. A strong downward cloud current causes what is known as a cloudburst, or downpour.

If a raindrop is any larger than a green pea, friction breaks it up as it falls. The maximum speed a raindrop can fall is 30 kilometres per hour (19 mph). Again, friction with the air prevents it from going any faster. As they fall, raindrops do not look like teardrops, as they are often drawn. Instead, they are shaped like tiny hamburger buns, with the flat side down.

 Freezing rain can severely damage trees, power lines, homes and buildings. When the ground is below freezing and the air is still above 0 degrees C (32°F), rain falls as drops that turn into ice on contact.

Usually, there is a layer of below-freezing air just above the ground. This supercools the raindrops, which means they are still liquid but a few degrees below freezing. Supercooled raindrops freeze almost immediately on contact with the ground.

10 THE BEAUTY OF RAINBOWS

- What colours are in a rainbow?
- What causes a rainbow?
- Is there a pot of gold at the end of a rainbow?

SUNLIGHT

△ A rainbow is a natural display of the colours of sunlight. Sunlight is a combination of red, orange, yellow, green, blue and violet light. In a rainbow, we see the colours spread out side by side in a beautiful arch. Red is always on the outside edge of a rainbow, violet on the inside.

When you look at a rainbow, you see the colours in separate positions because each colour is bent by a different amount as it passes through a raindrop. Some drops will be bending red light toward your eye, others green light, and so on.

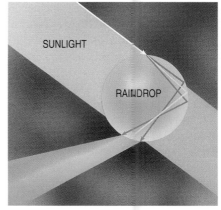

SUNLIGHT

RAINDROP

 A raindrop acts like a tiny prism. Sunlight is bent as it passes through the drop, just as it is with a glass prism. The bending spreads the light into its colours. Sunlight reflects off the back of the drop, causing the light to emerge in almost the same direction it entered.

Rainbows are only seen when the sun is behind the observer and falling water is in front. The water can be from a storm cloud, a lawn sprinkler or the mist above a waterfall. In all cases, water droplets are the cause.

A rainbow can be seen only when the sun is low in the sky. If the sun is high, the rainbow will be low or invisible because of the angle that raindrops bend sunlight.

You can make a backyard rainbow by using the fine spray of a garden hose. With your back to the sun when it is low in the sky, turn the nozzle until the spray is at its widest angle. A rainbow will appear in the water. It is easiest to see against a dark, shadowed background. Each person sees his or her own rainbow from different drops of water.

If you happen to see an especially bright rainbow in the sky, look for a second, dimmer one outside the main bow. The colours in the second one are in reverse order.

Rainbows are curved because they are part of a circle. To see the complete rainbow, you have to be in an airplane looking down on a rainstorm. That is why it is safe to say there is a pot of gold at the end of the rainbow—there is no end!

25

11 SNOW AND SNOWSTORMS

- What is snow?
- How does a snowflake form?
- Are there different types of snow?
- Do snowflakes have six points?
- Why is it quiet just after a fall of fluffy snow?

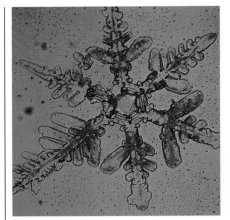

△ The two things most often said about snowflakes are that no two are alike and that they all have six points. While it is true that no two flakes are *exactly* alike, many do not have six points, or six sides. However, the largest snowflakes are the ones most likely to look like beautiful six-pointed stars.

Why six? This natural shape is caused by the way tiny crystals of ice attach to each other as the snowflake forms in the cloud. The crystals fit together like building blocks. It takes millions of ice crystals to create one flake.

Depending on the temperature and the amount of water vapour available during formation, snowflakes can also look like tiny bricks, tubes, needles or a multitude of other shapes, especially if they partially melt and refreeze on the way to the ground. There is tremendous variation in size too. The biggest fluffy flakes are actually groups of snowflakes that stick together on the way down. Merged flakes can be the size of a dime. Flakes that melt partially as they fall are just tiny blobs of ice the size of a ball in a ball-point pen when they reach the ground.

Generally, the largest flakes fall when the temperature is near freezing and there is little wind. Snowflakes falling on very cold days are always small, mainly because very cold air holds so little moisture that there is not much available for snow formation.

△ Wind can sculpt beautiful patterns in snow. The wind changes snow as it blows it around. Large flakes are smashed into other flakes and broken apart. Soon, there are no flakes left, just tiny ice particles. Since the small particles fit more closely together, the snow is made denser by the wind. Of course, it also blows it into drifts.

▷ Most of the five billion people on planet Earth seldom see snow because they live in areas of arid or tropical climate. But for the rest of us, winter—and snow—is a fact of life.

Surprisingly, regions in southern Canada and the northern United States receive more snow in a year than falls at the North Pole. The high Arctic near the North Pole is actually a frigid desert. But the little snow that does fall stays.

Snow is basically ice crystals and thousands of air spaces that make it light and fluffy, instead of hard and dense like ice cubes. On average, snow is 10 times lighter than water. Fill a pail with snow, bring it indoors, and let it melt to measure the water equivalent. The usual ratio of snow to water is 10 to 1. Thus 250 millimetres (10 inches) of snow equals 25 millimetres (1 inch) of water. Or think of it this way: if 25 millimetres of rain came down as snow instead, there would be 250 millimetres of snow on the ground.

The water-equivalent ratio can vary considerably. Very wet snow—"packing" snow ideal for making snowballs and snowmen—has a ratio of about 6 to 1, whereas extremely powdery snow can have a ratio of up to 30 to 1.

The ratio changes over time too. As snow lies on the ground, it becomes compacted by its own weight and by melting. Direct sunlight can melt snow even if the air temperature is below freezing. When this is the case, slow melting occurs near the ground. In Arctic regions, the snow does not melt but slowly turns into rock-hard ice as it is pressed down by more snow on top.

Have you ever noticed how quiet it is right after a fall of fluffy snow? Freshly fallen snow imparts a hush to the landscape. It absorbs sound. Fluffy snow is especially good because sound waves are trapped in the tiny openings between flakes. Large snowflakes tend to sit atop one another to create the openings that make snow nature's perfect sound absorber.

Snow is also excellent insulation because the fibrous structure of the crystals contains tremendous numbers of tiny openings that trap air. Snow crystals are nature's counterpart to modern home-insulation materials. In extreme cases, people lost in the bush in winter have buried themselves in snow to keep from freezing.

The heaviest snowfalls occur when the temperature is just a few degrees below freezing. (The colder the air, the less moisture it holds, so most snow is formed during mild winter conditions.) This is a typical winter condition across the middle of North America. Major cities such as Toronto, Montreal, Chicago and Detroit receive from 60 to 150 centimetres (2 to 5 feet) of the white stuff in a normal winter season. That is an average of 100 million tons of snow on each city. Farther south, cities like San Francisco, Houston and Atlanta seldom experience paralyzing snowstorms.

People who live near the Great Lakes are familiar with heavy snowfalls due to the lake effect. On their journey across the lakes, nimbostratus clouds gather additional moisture that is always evaporating from the water (the Great Lakes rarely freeze over completely). The extra cargo of moisture is quickly dumped as snow once the cooler land is encountered.

When vicious winds unite with the lake effect, the consequences are awesome. A January 1977 storm with 100-kilometre-per-hour (60 mph) winds lashed the city of Buffalo at the eastern end of Lake Erie. The snow was whipped into drifts as deep as 7.5 metres (25 feet).

29

12 THUNDERSTORMS AND LIGHTNING

- What causes lightning?
- Can lightning occur in winter?
- Can lightning strike twice in the same place?
- Does lightning go from the cloud to the ground or from the ground to the cloud?
- What causes thunder?
- What is the rule for determining the distance to lightning?

Thunderstorms usually erupt at the end of a hot summer day when cumulonimbus clouds are fully developed. Giant cumulonimbus thunderheads can measure more than 20 kilometres (12 miles) from top to bottom. Vertical air currents carrying moisture, water droplets and ice crystals within the huge clouds can create positive and negative electrical charges in the cloud. The electrical charges are released by lightning flashing either within the cloud or between the cloud and the ground. Both types are seen in this photograph. And, yes, lightning can occur during a snowstorm in winter and can strike in the same place more than once.

Lightning looks as if it has enormous power—and it does. Because there is always a thunderstorm somewhere, about 100 lightning bolts strike Earth every second. Scientists estimate that this yields a continuous power output of four billion kilowatts—enough to supply all of Canada's energy needs.

When lightning hits a tree, the moisture in the wood instantly expands and turns to steam, causing the tree to explode and send pieces flying in all directions. Even though some bolts are forked, usually only one prong actually completes the discharge.

Sometimes during a thunderstorm, only sudden glows in the clouds are seen, rather than lightning bolts. This is often called heat or sheet lightning. It is not a special type of lightning. Instead, the lightning is simply blocked from view by the clouds so that a pulse of light is seen but the bolt itself is hidden.

When a storm cloud discharges electricity to Earth, a small "leader stroke" zigzags toward the ground in about one one-hundredth of a second. Once it touches something—a tree, a pole, a barn or the ground— the electrical circuit is complete and a vast surge of electricity, the lightning bolt, roars up the path established by the leader stroke. Thus the lightning stroke generally goes from the highest object on the ground up to the cloud. It looks as if it happens the other way around—from the cloud to the ground. The human eye cannot follow the formation of a lightning stroke because it shoots skyward in about one-millionth of a second.

What the eye sees is the electrical discharge of as much as 100 million volts of electricity, which heats the surrounding air to a temperature five times hotter than the surface of the sun— 30,000 degrees C (54,000°F). A lightning bolt is just a few centimetres wide. Thunder is the sound caused by the instantaneous heating and expansion of air along the lightning bolt. The variety of thunder sounds is caused by the different shape of each bolt.

You see the lightning first and hear the thunder later because light travels about a million times faster than sound. A lightning flash is seen from a distance as soon as it occurs, but the thunder from the same flash takes a few seconds to reach the observer. By counting the number of seconds between the flash and the thunder and dividing by three, you can calculate the lightning's distance from you in kilometres (divide by five for miles).

During a violent lightning display such as the one in this time-exposure photograph, many objects could be hit— trees, lakes, barns, houses and swimming pools. Houses and cars are generally safe places for people to be during a thunderstorm. Water, open fields and solitary trees are not.

13 AIR CURRENTS

- What causes wind?
- Why does a hot day feel cooler when there is a breeze?
- Why does a cold day feel colder when the wind blows?
- Why do winds usually die down in the evening?
- Are winds stronger or weaker high in the sky?
- What is the jet stream?

△ Winds are currents in the atmosphere. Because the atmosphere is an ocean of air—and we live in that ocean—we feel the currents as winds.

In summer, winds feel cool. On a hot day, a breeze is welcome and refreshing. That is because wind increases evaporation by passing more air over damp surfaces. Wet clothes hung outside dry faster on a windy day. Our skin is not exactly a wet surface, but when sweat evaporates quickly in the wind, we feel cool.

In winter, cold air passing over a warm object carries away heat much faster than if there is no wind. The difference is called the windchill factor. For example, on a winter day when the temperature is minus 24 degrees C (-11°F), a 50-kilometre-per-hour (30 mph) wind will make it feel like minus 50 degrees C (-58°F). At that temperature, exposed skin freezes in one minute.

For the majority of us, who notice the winds only on the way to work or school, there are a few things to watch for. Clouds move with the wind. It takes less than a minute to tell which way clouds are moving. Stand still and observe cloud motion in relation to the fixed objects around you (trees, buildings, lamp posts). If the wind direction at ground level is different from the direction the clouds are moving, a change in the weather is likely in less than a day because a mass of cool air is probably wedging beneath the warmer air above.

Once you begin to notice the wind, you will find that on sunny days, the wind almost always dies down at sunset. That is because heat from the sun generates currents. When the sun is gone, so is the heat source, and things usually calm down. This rule does not hold on heavily overcast or stormy days because the clouds do not let sunlight reach the ground.

The atmosphere moves because air heated by sunlight rises, then cools and descends, only to be reheated again. This never happens smoothly, with a great mass of air going up and down like an elevator. Rather, parcels of air of widely varying sizes move at different speeds, and a general churning effect results. The process keeps the air in motion, with currents of air flowing from place to place—wind.

Because it is hottest near the equator and coldest at the poles, there is a general system of air circulation over the whole Earth that causes prevailing winds at any particular location. For example, in southern Canada, the northern United States and most of Europe, the strongest winds come from the west. In Hawaii and northern South America, the prevailing winds are from the east.

Actually, wind circulation over the planet is very complex. Powerful computers are needed to keep track of what happens each day. Some parts of the world are windier than others. For example, mountaintops are windier than valleys. Seashore towns and villages experience higher winds, on average, than inland communities.

The strongest winds—the jet stream— blow from 100 to 200 kilometres per hour (60-120 mph). They are 10 to 15 kilometres (6-9 miles) above sea level. Airplanes flying from North America to Europe often get a boost from the jet stream, which flows in the same west-to-east direction.

14 WEATHER LORE

Weather forecasting is a modern science. Years ago (before 1930), predicting tomorrow's weather was little more than a guessing game. From experience, farmers and fishermen learned what to watch for. They made up easy-to-remember rhymes and sayings about ways to predict weather. Some of their notions are still with us today. A few of them are based on scientifically valid concepts. But most are wrong as often as they are right.

Each February 2, we are reminded faithfully about the story of the groundhog and his shadow. If the groundhog comes out of his burrow at noon on that day and sees his shadow, we are in for six more weeks of winter. If not, according to the legend, mild weather can be expected. Weather records have proved the groundhog wrong many, many times.

However, some animal behaviour is a good predictor for the weather, at least during the next few hours. Some farm animals know when a storm is coming and will seek shelter; others become agitated and restless. Ants and spiders scurry at top speed to complete their tasks. Bees return to the hive.

Today, most weather lore is simply quaint —not very useful in modern life, but fun.

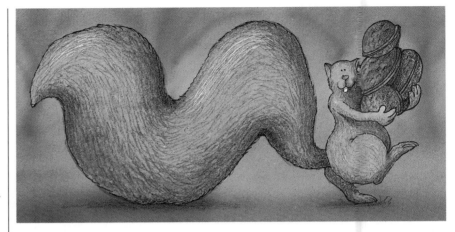

 "Squirrel's tail fluffy, winter will be blustery" is one of hundreds of weather rhymes about animals, birds and insects. Another supposed predictor is the woolly caterpillar: if the black band on the caterpillar's back is wide in autumn, the winter will be severe. Scientists say, however, that neither squirrels' tails nor caterpillars' backs predict winter weather.

"A coming storm your shooting corns will presage" is one of many old sayings suggesting that aches and pains are worse when a storm is approaching. This may be true. Some aches could be aggravated because humidity is maximum and atmospheric pressure minimum just before and during a storm.

"If the sun red should set, the next day surely will be wet." Well, it depends on where you live because the following rhyme says the opposite: "Red sky in the morning, sailors take warning; red sky at night, sailors' delight."

△ "If woolly fleeces spread the heavenly way, be sure no rain disturbs the day." This rhyme refers to a sky of small puffy cumulus clouds. It tells us that no rain will fall that day. This would be true most of the time, but a cumulonimbus cloud can develop, bringing thunderstorms.

◁ Another legend states that rain will arrive soon if frogs croak louder than usual. This is probably true, because frogs are most active in damp weather.

△ "It's raining cats and dogs" is a very old expression. In mythology, cats were associated with rain and dogs with wind. So a windy rainstorm was called a cat-and-dog storm.

15 SKY COLOUR

- Why is the sky blue?
- Is the sky on other planets blue?
- What causes colours in the sky?

We see the brilliant blue sky so often that we take its colour for granted. Very few people ever think about *why* the sky is blue. But skies are not blue on every world. The sky on Mars is pink. On the moon, it is black. Venus has a yellow sky.

Sunlight is composed of a spectrum of colours, the same colours that are seen in a rainbow. When sunlight enters the Earth's atmosphere, it encounters the atoms and molecules of air. All of the colours except blue travel straight to the surface, but blue light tends to bounce off the atoms and molecules that make up the atmosphere. Because the blue bounces around, it eventually reaches our eyes from all parts of the sky, not just straight from the sun, as the other colours do. Therefore, the sky looks blue. The sun appears yellow because yellow is what results when all of the colours except blue are combined. From space, astronauts say, the sun appears white, which is what it would look like to us if the blue were not removed and scattered before reaching our eyes.

On the moon, there is no air, so the sky is black. Dust in the Martian atmosphere makes the sky a pale orange or pink. Clouds on Venus tint the sky yellow.

The sky on Earth can be deep blue or pale blue or a washed-out light blue. The shade of blue depends on the amount of water vapour and dust particles in the air. Water vapour and dust scatter all colours of light, not just blue, and this causes a lightening of the blue background.

Very red skies after sunset can be seen for years following the eruption of a large volcano that has spewed huge amounts of dust into the atmosphere. Winds carry the dust around the globe. Volcanic dust scatters all colours of sunlight, blue the most and red the least. After sunset, all colours but red are blocked by the scattering, so red is the only colour that gets through to the observer.

This photograph was taken in Ontario in April 1982, one year after the eruption of the Mexican volcano El Chichon. The red sunsets continued for months, just as they did after a similar eruption of Mount Agung, in Indonesia, in 1963.

About half an hour after sunset, the sky is dark overhead but bright where the sun went down. Sunlight is not going around a corner but is scattered high in the air by dust so that the sky is still bright while the land has become dark. The air was very clear and contained little dust when this photograph was taken. The glow from scattered light spreads farther above the horizon the more dust there is in the atmosphere.

16 HAZE AND HUMIDITY

- Why is the sky sometimes clear and sometimes hazy?
- What causes haze?
- Why is haze most obvious at the horizon?

△ From day to day, there is a vast difference in the clarity of the atmosphere. Sometimes, the air is so clear that distant hills are easily visible. At other times, a haze makes the same distant objects dull or invisible. The haze has both natural and human causes.

Most haze is simply moisture in the air, but dust can be an important cause too. Dust comes from volcanoes and forest fires, and it can be picked up by wind blowing over desert and dry soil. Some haze is caused by smoke from factories and car exhaust gases. Haze created by human activity is usually called air pollution.

In this scenic panorama, photographed in Jasper National Park, in the Canadian Rocky Mountains, the atmosphere is about as clear as it can possibly be. In the distance, there is hardly a hint of any haze to dim the view of the mountains. That is because the air itself is clean and contains little moisture.

▷ Haze is caused by water vapour and tiny dust- or smaller-sized particles in the air. The water vapour and particles block part of the light from distant objects and scatter the light as well. We see a dull grey or bluish image, and details are difficult to make out. The greater the distance, the more the scene is dimmed, because the more air there is between us and a distant object, the more haze we must look through to see it. The photograph shows the effect very well, because the most distant hills are barely visible.

On days when the far horizon is hazy, the air has much more water vapour in it than when the same horizon is clear. Hazy days have high humidity, and clear days have low humidity. Hot days with high humidity are muggy and uncomfortable because human perspiration evaporates slowly into air that is already well supplied with water vapour.

17 FOG, MIST AND SMOG

- What are the differences between fog, mist and smog?

- Why does fog usually form at night?

- Why does smog develop over cities?

△ Fog is made up of droplets of water so small that they remain suspended in the air, just as they do inside a cloud. Fog can therefore be thought of as a cloud at ground level.

Most clouds form as air rises and cools, but fog emerges when the air at ground level cools at night and the water vapour in it turns into droplets. This usually happens in valleys where the air is still and over swamps and marshy regions where there is extra moisture available. Fogs generally occur on nights when the sky is clear. A cloud layer acts like an insulating cover, keeping air warmed during the day near the ground and preventing fog formation. Fog often becomes thickest around dawn, when the temperature is lowest and winds seldom stir. Then, sunlight begins to heat the atmosphere and the fog disappears as droplets turn to water vapour.

A second type of fog, called sea fog, is caused by air passing over cold ocean currents and then moving over land. It can last for many hours, even after the sun is up. Such fogs are often found in Newfoundland, San Francisco and parts of England.

▷ Mist is commonly seen in the early morning over or near a pond, river or small lake. During the night, the air and ground become colder than the water. As water vapour rises from the water, it meets the cooler air and condenses into drops that are too small to fall. The droplets stay suspended in the air, creating a miniature cloud, until sunlight warms the air and the droplets become vapour again. Mist never completely blocks the view to objects a few hundred metres away.

△ Smog looks like mist or fog over a city, but it is quite different. Smog is made by people. The word smog is a contraction of *smoke* and *fog*. But in today's cities, smog is more than just these two components. It is a combination of car exhaust, dust, water droplets and factory pollution that hangs over a city on calm days regardless of the humidity.

Smog is at its worst when there is no wind day after day and a layer of warm air above the city traps the air below so that there is little circulation—a condition called an inversion. The smog then builds into an ugly cloud that can become a health hazard. Fortunately, many cities do not often experience inversions. And strict pollution controls in recent years have helped prevent smog from becoming worse.

18 DEW AND FROST

- ■ Why does dew form?
- ■ What causes the patterns in frost?
- ■ Why is dew heaviest in the fall?

At night, grass and tree leaves cool more rapidly than the air. When it is clear and calm, water vapour in the air just above the grass and trees condenses on the plants in droplets (because cool air holds less water vapour than warm air). The droplets eventually collect into larger drops and soak everything exposed to outside air. Plants become wet first, then metal objects and wood and, finally, concrete and pavement, since they cool more slowly. If there is a wind, dew will not form because of the stirring of the air.

The largest amount of dew can be found at dawn on a clear day. If there have been clouds in the sky overnight, the dew will not be heavy because clouds act like a blanket and keep the temperature from dropping enough for maximum moisture condensation. After sunrise, the temperature goes up and the dewdrops begin to evaporate, but the process can take hours after a heavy dew.

The heaviest dews occur in autumn, when the nights are long and the air contains ample moisture.

If the air temperature drops below freezing on a still, cloudless night, the dew freezes and any additional moisture is deposited as frost. When moisture from the air freezes on window-

panes, it creates beautiful patterns called hoarfrost. The designs sometimes look like graceful leaves or feathers. Frost patterns are the result of the crystal structure of ice—the same thing that causes snowflakes. Scratches on windows or particles of dirt act as condensation nuclei and are the first things the crystals grow on. The shapes emerge as crystals grow away from these starting points.

Sometimes the air cools and a fog forms. If the temperature continues to drop, however, the fog freezes on everything it touches, producing a spectacular frosty wonderland at daybreak, as seen in this photograph.

In areas of very dry climate where it almost never rains, some desert plants and shrubs survive only by absorbing water from dew and frost at night.

43

19 | CLIMATE ZONES

- Why does Earth have wind-circulation belts?
- How does wind circulation affect weather?
- What is the difference between weather and climate?
- What are the main climate zones in North America?

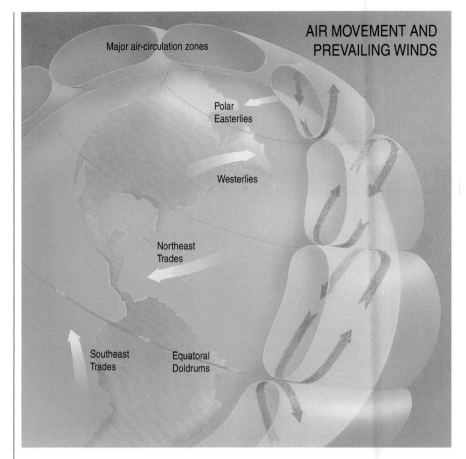

AIR MOVEMENT AND PREVAILING WINDS

Major air-circulation zones

Polar Easterlies

Westerlies

Northeast Trades

Southeast Trades

Equatoral Doldrums

△ An old saying is: the winds bring the weather. But there is more to it than that. The sun is the real "engine." Sunlight heats the ground, the ground heats the atmosphere, and the rising and descending currents in the atmosphere create three major wind-circulation belts in each hemisphere. The situation remains fairly constant from year to year, which is why annual weather patterns tend to be approximately the same at any specific place on Earth.

▷ Climate means the average or typical weather conditions during the year for any particular part of our planet. For example, the climate for Florida is very hot in summer and mild in winter, with frequent precipitation throughout the year. These conditions are known as humid subtropical climate. The map shows the main climate zones—10 regions of similar climate—in Canada and the United States.

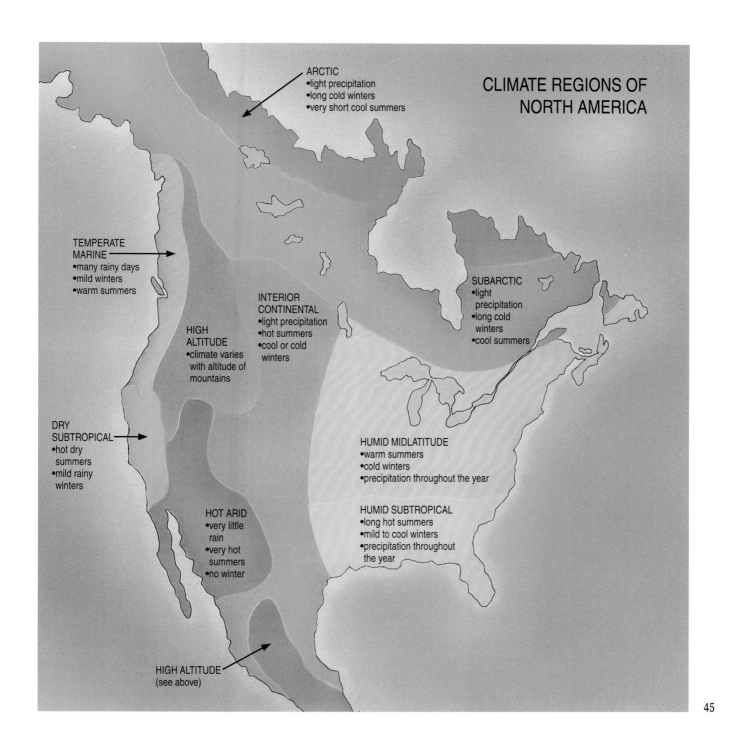

CLIMATE REGIONS OF NORTH AMERICA

ARCTIC
•light precipitation
•long cold winters
•very short cool summers

TEMPERATE MARINE
•many rainy days
•mild winters
•warm summers

SUBARCTIC
•light precipitation
•long cold winters
•cool summers

INTERIOR CONTINENTAL
•light precipitation
•hot summers
•cool or cold winters

HIGH ALTITUDE
•climate varies with altitude of mountains

DRY SUBTROPICAL
•hot dry summers
•mild rainy winters

HUMID MIDLATITUDE
•warm summers
•cold winters
•precipitation throughout the year

HUMID SUBTROPICAL
•long hot summers
•mild to cool winters
•precipitation throughout the year

HOT ARID
•very little rain
•very hot summers
•no winter

HIGH ALTITUDE
(see above)

45

- What is a sundog?
- What causes a pillar of light at sunrise or sunset?
- Why does a halo sometimes appear around the sun or the moon?
- Why are haloes usually seen in cool weather?

△ This beautiful splash of light, called a parhelion, is more commonly known as a mock sun or a sundog.

A sundog appears well to the right or left of the sun and only when the sun is fairly low in the sky. Cirrus or cirrostratus clouds must also be present because they are made of the right kind of ice crystals: six-sided and shaped like short pencils. The little ice pencils sometimes float straight up and down in the air. When they do, they refract, or bend, sunlight in just the right way to cause a sundog.

If the sundog is especially bright, it will be coloured—red on the side closest to the sun, blue on the far edge and yellow in the middle. Sundogs are brightest in winter because the tiny pencil crystals are more common then, but they can be seen every season. Two sundogs are seen in the photograph on the facing page.

△ At sunrise or sunset, a brilliant pillar of light, like a coloured searchlight beam, is sometimes seen directly above the sun. Called a sun pillar or a solar pillar, it is caused by sunlight reflecting off tiny ice crystals, usually at the same altitude as cirrus and cirrostratus clouds.

Only a certain type of ice crystal will work—it is shaped like a tiny six-sided

flying saucer and floats horizontally if the air is not turbulent. The crystals act like billions of miniature mirrors reflecting sunlight off their flat undersides. The pillar effect occurs because only the crystals directly above the sun are at the right angle to reflect sunlight to our eyes.

Sometimes, in winter during foggy weather, ice crystals form in the air just above the ground. Under these conditions, pillars can be seen above street lamps.

△ Although solar haloes are quite common, few people have seen one because the sky is generally very bright when a halo is present. The brightness is due to cirrostratus clouds, which cause the halo. Cirrostratus are thin, milky sheets of cloud that veil the sky but allow the sun to shine through. They are made of ice crystals that refract sunlight at a specific angle and thereby create the halo effect.

Solar haloes are most often seen in cooler weather, when conditions are ideal for the formation of the best type of cirrostratus clouds. Do not look close to the sun for a halo. Haloes are big, and besides, staring at the sun can damage your eyes. The best plan is to position yourself so that the sun is blocked from view by a building.

The same type of halo can be seen around the full moon. There is an old saying: "Ring around the moon, rain will come soon." It is only half right, though. Statistically, about half of the time a halo is visible, rain or snow follows a day or two later.

47

21 AIRPLANE WINDOW VIEW

- Why does the sky appear deep blue or purple from an airliner?

- Why are clouds almost blindingly white outside an airplane window?

- What causes the white trail behind a jet plane?

Some of the best cloud watching can be done from an airplane. Sit on the shady side of the plane to keep the glare of sunlight off the window. That is the right side when flying west, the left side when flying east. At airplane cruising altitude, usually about eight kilometres (5 miles), the atmosphere is much thinner than it is at the Earth's surface. Less sunlight is scattered by the air, and the sky looks deep blue or purple. If the plane could fly higher, the air would be too thin to scatter sunlight, and the sky would be black.

The tops of clouds look different than the bottoms we see from our normal ground-level view. Notice how bright and pure white the tops of most clouds are from the air. On the ground, we usually see the un-illuminated side of clouds—the side not directly in sunlight. But from an airplane, the full splendour of sunlit clouds is visible.

Cumulus clouds, like those in this photograph, are the most impressive. They have flat bottoms but billowy tops, like great heaps of cotton. From a distance, their edges are sharp and distinct, but if the plane passes close to them, they appear like mist, which of course is what they are.

spewed out the back of the engine is hot and contains moisture (water vapour) from the burning of jet fuel. The water vapour freezes in one or two seconds, forming a trail of ice-crystal cloud, just like a natural cirrus cloud. The short space between the jet and the contrail is due to the brief time it takes the moisture to freeze.

If a contrail vanishes rapidly, the air is dry and fair weather is likely to continue. Contrails can last for hours, however, and may even trigger the formation of more extensive natural cirrus clouds.

From a seat on the shady side of an airplane, you might be able to see what is known as the glory ringing the plane's shadow. It is a pale, circular rainbow caused by the scattering and reflection of sunlight from water droplets in clouds below the plane. However, the glory is not a true rainbow and is usually very small, so you have to watch carefully.

Trails left behind jet planes are called contrails. Contrails are cirrus clouds made by humans. A jet engine pulls in cold, dry air at high altitude. The air

22 SUNSET AND SUN SHAPE

- Why are sunsets red?
- How long does it take for the sun to set?
- Why doesn't it get dark right after sunset?
- Why does the sun sometimes appear oval-shaped at sunset?

△ Sunsets are red because all the other colours in sunlight are scattered and absorbed by atoms, molecules and dust particles in the air. Only red and a bit of orange remain. Those colours are scattered too, but just enough to spread a rich red or deep orange glow over the sky.

Clouds add to the beauty of a sunset, but only if they are high enough not to block the sun. The sun must be able to peek through and illuminate the underside of the clouds. That makes a majestic sunset.

The sky does not become dark immediately after the sun goes down because sunlight continues to illuminate the air overhead. Even an hour after sunset, there is still a twilight glow in the west due to sunlight on the upper atmosphere.

▷ Under normal circumstances, never look directly at the sun. Doing so can cause permanent damage to the eyes. Sometimes, though, at sunset, when the sun is a great orange or red ball on the horizon, like a giant pumpkin, it is so dim that it can be safely viewed. At such times, the sun appears to be not round but oval—wider than it is high.

When sunlight has to pass through a great amount of atmosphere, as it does when it skims the horizon at sunset or sunrise, the air bends the light rays, and the sun has a squashed appearance. The same effect, called refraction, makes people's legs look shorter when they are standing in a swimming pool.

It does not take the sun long to set. From the time the first edge touches the horizon to the moment the last sliver disappears is usually less than two minutes.

23 EARTH MOTIONS

- What causes the seasons?
- Why are the Earth's motions in space so important?
- When is Earth closest to the sun?

◁ We live on a moving planet. It has two main motions that have very noticeable effects on our daily lives. Motion number one is rotation, the Earth's 24-hour spin on its axis that causes day and night. Motion number two is revolution, the Earth's yearlong path, or orbit, about the sun. During the annual orbit, one hemisphere is angled slightly toward the sun; six months later, the other hemisphere is. This produces the cycle of the seasons.

Each season has its distinctive temperature range—hot in summer, cool to warm in spring and fall, cold in winter. In summer, when the northern hemisphere is angled toward the sun, two things happen to produce hot weather. First, the days are 13 to 15 hours long and the nights 9 to 11 hours. In winter, the opposite is true. Second, because the sun is high in the sky in summer, the rays of sunlight shine almost straight down and heat the ground much more effectively than in winter, when the sun is low and its rays hit Earth at a shallow angle.

The hottest days of summer occur one to two months after the longest amount of daylight in late June. Heat builds up in lakes, rocks and soil, reaching a maximum in July and August. Similarly, the coldest weather is in January and February, not in December when the nights are longest. (The cycle is exactly the reverse in the southern hemisphere.)

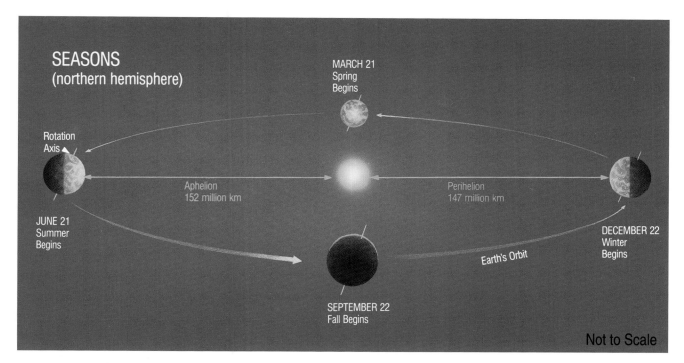

SEASONS
(northern hemisphere)

MARCH 21
Spring
Begins

Rotation
Axis

Aphelion
152 million km

Perihelion
147 million km

JUNE 21
Summer
Begins

DECEMBER 22
Winter
Begins

Earth's Orbit

SEPTEMBER 22
Fall Begins

Not to Scale

The cycle of the seasons regulates the basic weather patterns all over the world. It is a rhythm that governs much of our lives.

The tilt of the Earth's axis causes the seasons. The axis is tipped by 23½ degrees. The diagram shows how first the North Pole and then the South Pole are angled toward the sun.

The Earth's average distance from the sun is 150 million kilometres (93 million miles). Its orbit around the sun is 942 million kilometres (585 million miles) long. That means Earth travels nearly 2,600,000 kilometres (1,600,000 miles) a day.

The Earth's orbit is not a perfect circle. It is an ellipse—an egg shape, or oval. Earth is five million kilometres (3 million miles) closer to the sun in January than it is in June. However, the difference does not amount to much. In the northern hemisphere, we are only one Celsius degree (2 Fahrenheit degrees) warmer in winter and cooler in summer than we would be if the Earth's orbit were circular. The closest point to the sun is called perihelion, the farthest aphelion.

Another minor effect of the elliptical orbit is that the seasons are not the same length. Spring in the northern hemisphere is 93 days, summer 93, autumn 90 and winter 89. Of course, these are the calendar seasons. The weather can be seasonable or unseasonable regardless of the calendar.

24 AURORA BOREALIS

- What causes an aurora?
- Can an aurora be seen at any particular time of year or hour of night?
- Are the colours seen in photographs visible to the eye?
- How high in the sky is an aurora?

A common misconception about auroras (also called northern lights) is that they are caused by sunlight reflecting off ice in the Arctic. That this is *not* the case has been known for 70 years. The sun is responsible in another way, though.

Atomic particles (protons and electrons) are constantly being released into space by the forces that churn up the gases on the sun's surface. Large clouds of them are emitted during violent storms on the sun. When the clouds reach Earth, they become trapped in the Earth's magnetic field and are drawn toward the north and south magnetic poles. As the material strikes the upper atmosphere, there is an interaction between the sun's particles and the atoms in our atmosphere. The interaction releases beautiful coloured light.

A spectacular aurora display like this occurs several times a year over most sections of Canada and Alaska and some parts of the northern United States. Although it looks as if it is happening at the same height as the clouds, an aurora is much higher, usually 160 kilometres (100 miles) or more above the Earth's surface.

The brightest auroras generally have the most intense colours. Green is the most common colour, caused when light is released by oxygen atoms. Red auroras have been mistaken for the glow from major fires. Fire trucks have been called more than once to extinguish what really was a red aurora.

Aurora displays can build into dramatic light shows that illuminate the entire sky, sometimes for hours, with glowing, pulsating curtains. At such times, some people say they hear swishing sounds. Scientists have no explanation.

▷ Small, green auroras like this one are typical of what can be seen several times a year from southern Canada or the northern United States. During such displays, occasional rays or fans of light extend overhead, but generally, the glow stays close to the northern horizon.

Auroras are seen in the northern part of the sky most often, but they can appear as pale bands crossing the whole sky from east to west. There is no way to predict when an aurora will occur or what it will look like. A huge eruption on the sun can cause a spectacular aurora, but a similar disturbance a few months later might generate no auroras at all. There is no season for auroras, either. They can occur at any time of year. But some years are better than others. The next peak of sunspot activity, in 1990, 1991 and 1992, should be the best time for auroras until the 21st century. However, the only way you can be sure to see an aurora is to go outside every night and take a look.

Don't expect to see colours in auroras as brilliant as they appear in these pictures. Photographic film is more sensitive to faint colours than the human eye. However, the colours are there if you watch carefully.

▷ A bright aurora can be seen from almost anywhere, but for the best view, a dark location in the country is needed. The light of the city can wash out the fainter auroral light.

An aurora seen in the northern hemisphere is called aurora borealis. A southern hemisphere aurora is known as aurora australis.

25 MOON PHASES

- Why does the moon have phases?

- Is there one side of the moon we never see?

- What are the dark patches on the moon?

- Are the shadings the same every night?

△ The moon is our nearest neighbour in space. It is also the only celestial object other than the sun bright enough to be easily visible through the blue of the daytime sky. To understand how far away the moon is from us, imagine a highway stretching all the way around Earth. Ten such highways placed end to end would span the distance from our planet to the moon. That is only a short hop compared with the distances to other objects in space. Because the moon is so close, it is seen as a world, rather than just a dot in the sky.

This photograph shows the most beautiful of all of the moon's phases, the bright crescent; the remaining part of the moon is faintly visible. In previous generations, people referred to this as "the old moon in the new moon's arms," but the expression is seldom heard any more. The faint lighting of the moon's night side is called earthshine.

Earth and the moon are huge globes of rock that have no light of their own. The sun shines on both worlds, producing day and night. The bright crescent seen in this photograph is part of the section of the moon that is in sunlight. If you could stand on that side of the moon, the sun would be in the sky and the ground would be sunlit.

Now, picture yourself standing on the darkened portion of the moon. The sun is not in the sky. It is night. But the landscape is faintly illuminated by sunlight reflecting off Earth—earthshine.

△ When the moon is a crescent in our sky, Earth is almost full and is near its brightest in the moon's sky. There is enough earthshine at such times to make the night side of the moon visible.

The best time to see earthshine is in late winter and early spring—February, March and April—when the crescent moon is farther above the horizon than it is in other seasons (from the northern hemisphere). Look at a calendar to determine the date of the new moon, then watch the western sky after sunset two to four days later. After four days, the crescent is too bright and earthshine too faint to be seen without binoculars or telescopes.

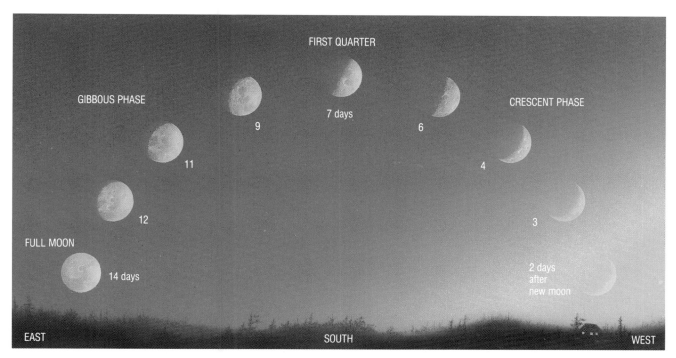

FIRST QUARTER

GIBBOUS PHASE

7 days

9

11

6

CRESCENT PHASE

4

12

3

FULL MOON

14 days

2 days
after
new moon

EAST   SOUTH   WEST

Of the many cycles in nature, the phases of the moon rank among the most important and easiest to observe. The moon is visible from anywhere on Earth almost every day of the year. Most of the time, it can be seen during part of the day and part of the night.

The moon orbits Earth at an average distance of 385,000 kilometres (239,000 miles). One complete orbit takes just under a month. During that period, the moon runs through all of its phases. The diagram on this page shows half of that cycle: the two weeks from thin crescent to full moon, as seen at dusk. Basically, we are showing half an orbit.

The phases change because we see different amounts of the daytime side of the moon as it circles Earth. Once in each orbit, the moon sweeps through the region be-

tween the sun and Earth. This is new moon, a phase that is not visible because the sun is seen in the same direction.

Two days after new moon, a thin crescent moon becomes visible low in the west after sunset. As the moon continues in its orbit, the phase increases. About a week after new moon, it is one-quarter of the way around its orbit and appears half illuminated. This is called first-quarter phase. First quarter is a confusing name because it does not refer to the moon's shape. It should have been called half moon.

A few days later, the moon advances to the three-quarter phase. This is called gibbous moon, which means humped. About two weeks after new moon, the moon reaches the point in its orbit where it is opposite the sun, and we see it entirely illuminated. This, of course, is full moon.

The full moon rises in the east as the sun sets in the west.

After full phase, the moon is below the horizon until after dusk and therefore is not shown in this illustration. During the two weeks following full moon, the phases are repeated in reverse order as the moon swings around to complete its orbit and the cycle begins again at new moon. The first half of the cycle, as shown here, can be seen in late afternoon and early evening. The half of the cycle after full moon is best seen around dawn or in early morning.

Centuries ago, long before the invention of the telescope, the moon was a magic lamp in the sky—at least that is what it seemed to be. Nobody knew why it shone brightly at night but dimly during the day, nor did they understand the nature of the dark and light areas on the moon's surface. But there is no doubt that people paid attention to the moon and its phases. They saw the phases just as we see them today, as shown in the three photographs on these two pages.

More than 30,000 years ago, early humans recorded lunar phases in bone carvings. Crops were planted according to the moon's phases, and the first calendars were based on the lunar month. About 4,000 years ago, observatories were built and staffed with astronomers who recorded the phases of the moon as well as the positions of the planets. About 150 B.C., the Greek astronomer Hipparchus determined the first reasonably accurate distance to the moon. Greek astronomers were fully aware of the moon's size and the fact that it orbits Earth. Aristotle even said the moon was a solid, rocky object illuminated by the sun. Today, the moon holds few mysteries for researchers.

The mystery may be gone, but the majesty remains. Look at the moon with binoculars. The detail visible is amazing, especially the craters in the light areas and the vast dark plains. If you ever have a chance to see the moon through a telescope, take a good long look. It is like peering out a spaceship window.

Some people say the dark and light splotches are arranged like a face—the Man in the Moon. The same pattern of features is visible month after month, year after year, because one side of the moon always faces Earth and the other side is turned away. Only the phase changes; the moon does not turn with respect to Earth.

What, then, is the dark side of the moon? It is not the side that faces away from Earth. The dark side is the nighttime side. These photographs show that most of the time, we are seeing part of the moon's daytime side and part of its nighttime side.

The full moon rises in the east around sunset and reaches its highest point in the sky about midnight. Although the full moon is dazzling when the sky is dark, it is really insignificant compared with direct sunlight. If the entire sky were filled with full moons, lined up side by side from horizon to horizon, the combined light would not equal one-quarter of the sun's brightness.

The gibbous moon is the phase most frequently seen in full daylight because it is well away from the sun's glare and there is lots of illuminated lunar surface reflecting sunlight. The gibbous moon is especially high in the eastern sky during the winter months.

26 HORIZON ILLUSION

- Why do the sun and moon appear larger when close to the horizon?

- Does the atmosphere act like a lens to magnify the size?

△ When the sun and the moon are close to the horizon, they look much larger than when they are high in the sky. Yet there is really no difference at all. Not a bit. The sun and the moon are exactly the same size whether they are close to the horizon or not. It is an optical illusion.

The sun is a huge, glowing ball when it sets, so it is difficult to accept the statement that the sun is no bigger near the horizon than when it is high in the sky. Because the moon is not as bright as the sun and is therefore easier to observe, the horizon illusion is even more obvious.

Could the atmosphere be bending the light near the horizon or magnifying the image? While it is true that the atmosphere does bend light, it does not magnify.

Is it possible that we are closer to the sun or the moon when they are near the horizon? No. Calculations show that the observer is slightly *farther away* from these bodies when they rise and set.

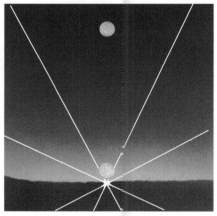

△ Nobody is exactly sure why the human mind insists on making the sun and moon so big at the horizon, but it is at least partly due to the illusion shown here. The upper ball looks smaller, yet if you measure the two, you will find that they are identical. The same thing seems to happen when we see the sun or the moon high and low in the sky.

Researchers have conducted experiments in a darkened room with two identical lighted balls, one straight ahead and one just about overhead. Almost everyone brought into the room said that the ball straight ahead was bigger.

Just as the balls in the illustration can be measured, it is easy to prove to yourself that the moon is always the same size. Take an aspirin tablet and hold it in front of the moon, with your arm out straight. The tablet neatly covers the moon no matter where it is in the sky.

When near the horizon, the moon is often tinted a gold or orange colour. This occurs for the same reason the sun has these colours at sunset. On rare occasions, the moon enters the Earth's shadow, creating a lunar eclipse. If it is a total eclipse, such as those visible from North America on August 17, 1989, December 9, 1992, and November 29, 1993, the moon is a deep coppery hue for about an hour while it is within the shadow.

Even rarer than a lunar eclipse is a blue moon. The blue colour results when a certain type of volcanic dust or smoke from extensive forest fires reaches the upper atmosphere. Conditions have to be just right. Few people ever see a blue moon, hence the expression "once in a blue moon."

Every three years or so, there are two full moons in a calendar month. News reports call the phenomenon a blue moon. However, most astronomers say that is not the real blue moon, but a modern substitute. True blue moons are very rare.

27 HURRICANES AND HAIL

- What are hurricanes?
- Where do they come from?
- How fast does a hurricane travel?
- Are hurricanes mostly wind, mostly rain or a mixture of both?
- Are hurricanes the same as typhoons?
- How does hail form?

△ Hurricanes are large, intense storms with powerful winds. To be classed as a hurricane, a storm's winds must exceed 117 kilometres per hour (73 mph). Because a hurricane is so vast—often larger than the province of New Brunswick or the state of Ohio—it can affect a huge region.

In this satellite photograph of Earth, there is a hurricane in the Gulf of Mexico, just above centre. It is the largest of several storms that look like pinwheels of cloud.

The heart of a hurricane is a great mass of cumulonimbus and nimbostratus clouds that have accumulated huge amounts of moisture evaporating from tropical seas.

Hurricanes always form over warm ocean water (hurricanes that develop over the Pacific Ocean are known as typhoons). At the fringe of a hurricane, the winds are gusty but not severe. Over several hours, however, as the centre of the hurricane approaches, the winds increase to furious levels. Ships at sea are in great danger, and giant waves crash into the shore if the hurricane is off the coast.

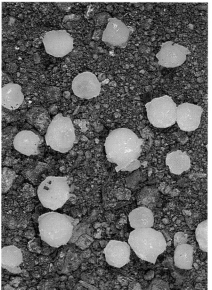

△ Hurricane Gladys is seen in this photograph taken by an astronaut from the Apollo 7 spacecraft in 1968. Hurricanes are alternately given male and female names to make them easy to refer to in weather reports and to distinguish one storm from another. The practice of naming hurricanes began in 1953, although only female names were used until 1979.

Hurricanes usually have a one-two punch: wind and rain, and lots of both. Winds can reach 200 kilometres per hour (125 mph), and enough rain can fall to cause major flooding. The states around the Gulf of Mexico are subjected to the greatest number of hurricanes. Usually, a hurricane loses its power if it stays over land, but not always. The big storms sometimes continue north, even as far as Canada.

Hurricanes move at speeds of up to 90 kilometres per hour (55 mph) and can therefore cover a lot of territory in a day. They seldom travel in a straight line, so weather forecasters have to rely on educated guesswork in predicting where a hurricane will be in 24 hours. Today, forecasters are aided by computers and detailed knowledge of where past hurricanes went.

△ Hail can fall during a hurricane, but it usually occurs during a severe summer thunderstorm. Inside a cumulonimbus cloud, wind currents can sweep tiny ice crystals back and forth, up and down. If the up-and-down currents are particularly severe, an ice crystal gets bigger and bigger as water droplets hit it and freeze layer upon layer, like an enlarging snowball. But hail is solid ice—heavy and dangerous.

Hailstones can range anywhere from the size of a pea to that of a baseball. The hailstones in this photograph are about the size of cherries. Hail falls to the ground at 160 kilometres per hour (100 mph) and does tremendous damage to crops and buildings. If you break a hailstone in half, you can see the layers, like those in an onion, each one formed as the piece was carried up by cloud currents to frigid altitudes, then down, then up again.

28 TORNADOES AND WATERSPOUTS

- Where and when do tornadoes and waterspouts occur?
- What causes them?
- Why are tornadoes so destructive?
- How fast are the winds in a tornado?

△ Tornadoes, sometimes referred to as twisters, are born inside giant cumulonimbus storm clouds. They are nature's most ferocious and terrifying storms. Since it takes special conditions for tornadoes to emerge, they are never seen in most parts of the world. Almost all tornadoes occur in central North America. Several hundred times each spring and summer, tornadoes are reported in a region known as Tornado Alley in Texas, Oklahoma, Nebraska, Kansas and Missouri. Smaller numbers of tornadoes occur in the states surrounding Tornado Alley (Wisconsin, Illinois, Iowa and Indiana), and once every few years, a large or destructive tornado hits Ontario or one of the Prairie Provinces.

Fierce, swirling winds during especially violent summer storms can create a funnel-shaped cloud that reaches from a black cumulonimbus cloud toward Earth—the dreaded tornado. People once thought that the tentaclelike cloud of a tornado acted like a vacuum-cleaner hose—sucking up objects that it passed over. Although there may be some vacuum effect, scientists think it is almost entirely the brute force of the tornado's winds—up to 500 kilometres per hour (300 mph)—that do such terrible damage. Trees are uprooted by the churning winds, and objects as big as houses can be thrown into the air.

Tornadoes can move across land as fast as a car on a freeway and destroy everything in their path. The average tornado is about 100 metres (330 feet) wide at the base. The very largest are wider than 10 city blocks.

△ The damage inflicted by tornadoes is spectacular and deadly. In Canada and the United States, more people have been killed by tornadoes than by floods or hurricanes. The safest place to be if a tornado hits is the southwest corner of a basement or underground structure.

◁ Waterspouts are related to tornadoes, but they often occur without the severe storms that accompany tornadoes. They are white, while land tornadoes are grey or black from swirling dust and dirt. Waterspouts are more common than tornadoes, but they are far less dangerous because their winds seldom exceed 80 kilometres per hour (50 mph). This photograph was taken from the shore of Lake Winnipeg.

29 WEATHER FORECASTING

- How accurate are weather forecasts?

- Are weather forecasts today any better than they were 20 or 30 years ago?

- How is the information for a forecast collected?

- What tools are used to gather weather information?

- Do television weather people make their own forecasts?

△ A weather forecast begins with the collection of information. This photograph shows a rain gauge, the black barrel in the foreground. Each day at the same time, the contents of the barrel are carefully dumped into a calibration cylinder, the thin device beside the barrel. The calibration cylinder allows a very precise measurement of the amount of rain in the last 24 hours. This information, along with other measurements, is sent by telex to a large weather-forecasting centre to help predict weather in the direction that the rain clouds are moving.

△ Thousands of standard weather-recording stations are scattered around the world. This one is located on a mountain in Washington State near the Canadian border. Some stations are in the middle of large cities, but most are in rural areas. Every airport has one too.

Each station has instruments to record the high and low temperatures for the day, humidity, barometric pressure, wind speed and direction and amounts of rain or snow. The white box contains some of these instruments. It is designed to keep direct sunlight out but to allow air in. The wind gauge is seen in the background.

△ This is a larger and more sophisticated weather station than those shown on the facing page. The men at bottom centre have just launched a weather balloon that is quickly rising into the sky. Hundreds of hydrogen-filled rubber balloons are sent aloft worldwide twice each day, at noon and at midnight Greenwich Mean Time. A box of instruments dangling on a string below the balloon measures the changes in temperature, pressure and humidity as the balloon rises. A small transmitter in the instrument box radios the readings back to the weather station.

After an hour or two, the balloon reaches an altitude of 25 to 30 kilometres (15-18 miles). Because of the low air pressure at this altitude, the balloon bursts. The instrument package drops to Earth with a small parachute to prevent damage or injury to anything on the ground. Most instrument packages are never found, and in any case, they are not used again.

As the balloon rises, its course is tracked by antennas at the station. Since the balloon is pushed along by the wind, the balloon's motion gives information on wind speed and direction in the atmosphere above the Earth's surface. The information is vital to meteorologists (weather scientists) when they prepare forecasts.

Information from weather-recording stations is sent to a major forecasting centre where meteorologists use modern high-speed computers as well as old-fashioned experience to predict what tomorrow's weather will be. In the last 20 years, satellite photographs have given meteorologists a look from above the atmosphere, a view they never had until the space age.

Radar is another weather tool. The screen shown at right is a radar display. Radar signals sent out from a weather station will reflect off rain or snow, producing "echoes" that are detected by a receiver at the station. Heavy rainfall produces the strongest radar echoes.

Using a combination of radar, satellite images and reports from weather stations, the meteorologist prepares a forecast for today, tomorrow and perhaps two or three days after that. Maps and written forecasts are then sent by telex to airports, highway departments, newspapers and television and radio stations.

The most familiar way that many of us receive weather information is through radio and television. On television, the weather person can use elaborate maps to show temperatures, regions of high and low pressure and the movement of warm

and cold fronts. High-pressure zones generally bring sunny weather; low pressure means clouds and possibly precipitation.

Unfortunately, the weather maps on television are on the screen for just a few seconds. If your local newspaper prints a similar map, examine it and compare the maps from day to day. This will give you an idea of the repeating weather patterns, the way weather systems generally move from west to east and which ones bring good and which bring foul weather.

The best part of the local television weather is the satellite photograph. It really shows what is going on. You can see great sweeping patterns of cloud and quickly determine whether a thick cloud deck is on its way from the west to where you live.

Most television and radio stations have one of their news announcers give the weather. The announcer uses a forecast prepared by a team of meteorologists at a major weather-forecasting centre. Only a few stations in the larger cities have a trained meteorologist on the staff as the weather person.

What about the forecasts in almanacs that are printed more than a year ahead? Despite what some people say, almanacs are wrong as often as they are right.

68

Few radio stations can afford a full-time staff meteorologist, yet they want to provide their listeners with the facts presented by a qualified person. In most cases, meteorologists employed by the government to produce forecasts do not have time to be on the telephone giving detailed forecasts to dozens of radio stations. The meteorologists heard on many stations work for private weather-forecasting companies that are paid for the service by a group of stations. This spreads the cost so that even small stations can give the impression that they have their own weather person.

The meteorologist in this photograph works for a company that provides forecasts for radio stations all over North America. Each city's forecast and other information is on a clipboard on the wall of his studio. He constantly updates the forecast to ensure that his customers have the latest information.

In our grandparents' day, before satellites and global networks of trained workers who gather and analyze weather data, forecasting involved more guesswork than today. Now, it is extremely rare if tomorrow's weather is not accurately forecast. Even two-day forecasts are right three times out of four. However, it is still very difficult for meteorologists to produce accurate forecasts more than two days in advance. The Earth's weather machine is far too complex and unpredictable for us to peek more than just a few days ahead.

CREDITS

p. 3 Courtney Milne, Miller Comstock Inc.
p. 4 © William H. Mullins
p. 6 Thomas Kitchin, Valan Photos
p. 7 (left) NASA Photo
p. 7 (right) illustration by John Bianchi
p. 8 Fred Whitehead, Earth Scenes
p. 9 illustration by John Bianchi
p. 11 illustration by John Bianchi
p. 12 (top) © N.R. Lightfoot, Miller Comstock Inc.
p. 12 (bottom) © Breck P. Kent, Earth Scenes
p. 13 © R.F. Head, Earth Scenes
p. 14 © Alister Ling
p. 15 (top) Cosmo Condina
p. 15 (bottom) P. Van Rhijn, Miller Comstock Inc.
p. 16 Courtney Milne, Miller Comstock Inc.
p. 17 Terence Dickinson
p. 18 J.R. Page, Valan Photos
p. 19 © Scott Rowed
p. 20 Robert C. Simpson, Valan Photos
p. 21 illustration by John Bianchi
p. 22 © W. Griebeling, Miller Comstock Inc.
p. 23 (left) © C.R. Potter, Miller Comstock Inc.
p. 23 (right) Martin Kuhnigk, Valan Photos
p. 24 illustrations by John Bianchi
p. 25 © Alan Dyer
p. 26 (left) © Harold V. Green, Valan Photos
p. 26 (right) Martin Kuhnigk, Valan Photos
p. 27 Thomas Kitchin, Valan Photos
p. 28 J.R. Page, Valan Photos
p. 29 (left) Jeff Foott, Valan Photos
p. 29 (right) © Charles Palek, Earth Scenes
p. 30 © John L. Pontier, Earth Scenes
p. 31 © Alan Dyer
p. 32 Kennon Cooke, Valan Photos

p. 33 F. Prazak, Miller Comstock Inc.
p. 34 illustrations by John Bianchi
p. 35 illustrations by John Bianchi
p. 36 © S.H. Suddes
p. 37 (top) Terence Dickinson
p. 37 (bottom) Courtney Milne, Miller Comstock Inc.
p. 38 © Roger N.J. Hostin, Miller Comstock Inc.
p. 39 © Klaus Brasch
p. 40 © W. Griebeling, Miller Comstock Inc.
p. 41 (top) © C.B. & D.W. Frith, Bruce Coleman Inc.
p. 41 (bottom) © W. Griebeling, Miller Comstock Inc.
p. 42 © Pam Hickman, Valan Photos
p. 43 (top) © Charles Palek, Earth Scenes
p. 43 (bottom left) © W.D. Griffin, Earth Scenes
p. 43 (bottom right) © Kenji Nagai, Miller Comstock Inc.
p. 44 illustration by John Bianchi
p. 45 illustration by John Bianchi
p. 46 (top) © Richard Kolar, Earth Scenes
p. 46 (bottom) © Alister Ling
p. 47 © Alan Dyer
p. 48 © Margot Conte, Earth Scenes
p. 49 (top) © Alan Dyer
p. 49 (bottom) Courtney Milne, Miller Comstock Inc.
p. 50 © Kozmik, Miller Comstock Inc.
p. 51 T. Mulligan, Miller Comstock Inc.
p. 52 (top to bottom) © Tom W. Hall, Miller Comstock Inc.; © Wayne Shiels, Four Winds Prairie Photography; © P. Van Rhijn, Miller Comstock Inc.; © Thomas Kitchin, Valan Photos
p. 53 illustration by John Bianchi
p. 54 © Terence Dickinson
p. 55 (top) Terence Dickinson
p. 55 (bottom) © Alan Dyer
p. 56 (left) © James Rouse
p. 56 (right) illustration by John Bianchi
p. 57 illustration by John Bianchi

p. 58 © Stephen J. Krasemann, Valan Photos
p. 59 (top) Halle Flygare, Valan Photos
p. 59 (bottom) J.R. Page, Valan Photos
p. 60 (top) © Stephen J. Krasemann, Valan Photos
p. 60 (bottom) illustration by John Bianchi
p. 61 © Wayne Shiels, Valan Photos
p. 62 NOAA Photo, Photo Researchers Inc.
p. 63 (left) NASA Photo, Photo Researchers Inc.
p. 63 (right) © Breck P. Kent, Earth Scenes
p. 64 © E.R. Degginger, Earth Scenes
p. 65 (top) © John A. Wood, Miller Comstock Inc.
p. 65 (bottom) John Scott Norquay
p. 66 (left) © Ray Nelson, Phototake Inc.
p. 66 (right) © Keith Gunnar, Bruce Coleman Inc.
p. 67 Joe Munroe, Photo Researchers Inc.
p. 68 (top) © Hank Morgan Photo Researchers Inc.
p. 68 (left) courtesy CJOH-TV
p. 68 (centre) courtesy CJOH-TV
p. 69 © Hank Morgan, Photo Researchers Inc.

FURTHER READING

Here are some well-illustrated introductory guides to the atmosphere and weather that are somewhat more detailed than the book you are now reading: *Weather* by Lehr, Burnett and Zim (Golden Press, New York, 1975); *Weather and Forecasting* by Dunlop and Wilson (Macmillan, New York, 1987); *Weather* by Atkinson and Gould (Mitchell Beazley, London, 1987); *The Skywatcher's Handbook* by Ronan, Dunlop and Jones (Crown, New York, 1985).

Each year, Environment Canada issues "The Canadian Weather Trivia Calendar," an entertaining and informative publication available in many bookstores or from the Canadian Government Publishing Centre, Supply and Services, Ottawa, Ontario K1A 0S9.

Weather lore and much other useful information can be found between the covers of *The Weather Book* by Reuben H. Hornstein (McClelland and Stewart, Toronto, 1980). Another excellent source of information on all aspects of weather and atmosphere is *The Sky Observer's Guidebook* by Charles E. Roth (Prentice-Hall, New York, 1986). *Atmosphere* and *Storm* are two beautifully illustrated and well-written books published by Time-Life (New York, 1983) in its Planet Earth series.

For the more advanced skywatcher, there is *A Field Guide to the Atmosphere* by Schaefer and Day (Houghton Mifflin, Boston, 1981) and *The Nature of Light and Colour in the Open Air* by M. Minnaert (Dover, New York, 1954).

The companion book to *Exploring the Sky by Day* is *Exploring the Night Sky* (Camden House, 1987).

INDEX